Beasley Library

Album of Whales

By TOM McGOWEN

Illustrated by ROD RUTH

RAND McNALLY & COMPANY
Chicago • New York • San Francisco

Text and illustrations reviewed and
authenticated by William P. Braker,
Director, John G. Shedd Aquarium,
Chicago, Illinois

Library of Congress Cataloging in Publication Data

McGowen, Tom.
  Album of whales.

  Includes index.
  SUMMARY: Introduces various types of whales,
dolphins, and porpoises.
  1. Cetacea—Juvenile literature. [1. Whales.
2. Dolphins. 3. Porpoises] I. Ruth, Rod.
II. Title.
QL737.C4M22   599.5   80-12904
ISBN 0-528-82287-X
ISBN 0-528-80070-1 (lib. bdg.)

First printing, 1980
Second printing, 1982

Contents

Whales—OUR GIANT RELATIVES IN THE SEA

WHALES ARE marvelous animals! Some of them are the very biggest creatures to have ever lived on the earth. That alone makes them special. But they have other special qualities, too. Some of them are the most intelligent animals in the sea—and may well be the most intelligent of *all* animals, next to humans. Some of them have special abilities that put our radar and sonar devices to shame. And some have a "song" that's more complex and thrilling to hear than the song of any bird!

It may surprise you, and please you, to learn that you're actually related to these fascinating creatures. You may wonder how an animal that lives in water and looks like a fish can be related to creatures that live on land and have arms and legs, but although whales do resemble fish—and many people mistakenly think they are fish—they're really much more like us than they are like any fish. Whales belong to the "family" of animals known as mammals, which is the family to which we also belong.

They may be armless, legless water animals now, but their ancestors were four-legged mammals that once lived on the land. The ancestors of fish, on the other hand, were always water creatures.

If you look at a fish's skeleton, you'll see that its front fins are simply fans made up of many long, very thin bones. But a whale's skeleton shows that inside a whale's front flippers there are leg bones and foot bones. And, farther back on a whale's skeleton there is a small hipbone, showing that the whale's ancestors once had back legs, too. As a matter of fact, unborn baby whales have four tiny, lumpy legs for a time. The front two become flippers, and the back two disappear before the baby is born.

Another way in which a whale is different from a fish is the way its tail is formed. A fish's tail fins are straight up and down, and to swim, the fish wiggles its tail from side to side. However, a whale's tail fins, called flukes, stick out sideways, and a whale moves its tail up and down.

Still another difference is that while no fish in any sea has any hair on its body, most whales do. Usually there are only a few stiff whiskers on a whale's chin, but they serve as a reminder that whales, indeed, belong to the hairy mammals—which include bears, monkeys, dogs, and all other such furry beasts—and not to the scaly fish.

There are other, even more significant differences. A fish is the kind of animal that's called *cold-blooded,* because its body is always the same temperature as the water in which the fish is swimming. That's why a live fish feels cold and clammy if you touch it. But a mammal is *warm-blooded.* Its body is always the same warm temperature, no matter how hot or cold the air or water around it may be. Whales are warm-blooded, and their body heat is kept from leaking out into the cold water by a thick layer of fat, called blubber, beneath the skin. Thus, even though fish and whales are shaped much alike on the outside, their inside "machinery" is very different.

The biggest difference of all between a mammal, such as a whale, and a fish is in the way it comes into the world and the food it eats as a baby. Most baby fish come out of their mother's body inside an egg. They grow and develop within the egg, then push their way out, or *hatch*. But most mammal babies stay inside their mother's body while they grow and develop. They come out, or are *born*, fully formed. The food of newly hatched baby fish is bits of plants or tiny animals they have to get for themselves. But the food of all baby mammals is a very special food that is furnished to them by their mother. That food is milk. It is made in the mother's body, and the baby sucks it out through openings called mammae. That word, *mammae*, is where the name *mammal* comes from.

So, although a whale may look like a cold-blooded, scaly fish that hatched from an egg, it's actually a warm-blooded, smooth-skinned animal that was born out of its mother's body and fed milk as a baby, just as you were. It's a mammal, just as you are—your relative in the sea!

Nature had to do a lot of changing to the bodies of whales' ancestors to enable them to live in the sea. One big change was to get rid of their legs and give them a streamlined, fish-shaped body. But there were other changes, too.

One other change enabled whales to stay underwater for long periods of time. All mammals can breathe only air, and that makes things difficult for a mammal that must dive down into water to get its food. It has to be able to stay down long enough

to locate the food and then to catch it. The longer an animal can remain underwater, the better its chance of getting something to eat. So, whales' ancestors had to become able to keep a great deal of air in their bodies in order to remain underwater for a long time. Now, some kinds of whales can stay underwater for as much as two hours! But, of course, a whale *must* come up to the surface eventually to poke its nostrils into the air and breathe. It's still an air-breathing mammal, and if it should somehow get trapped underwater, it will drown.

When a whale comes to the surface for fresh air, it has to let out the air it has been holding. It does this by *spouting*. Perhaps you've seen a picture of a whale with what looks like a cloudy geyser shooting up out of its head. The whale was spouting. A whale's spout is simply a cloud of moist air. Because a whale is a warm-blooded animal, the air it holds in its lungs when it is underwater gets hot and steamy. When the whale reaches the surface, it lets this hot breath shoot out of the *blowhole*, or nostrils, on top of its head. (Some whales have two nostrils, and some have only one.) The air over the water is far cooler than the air that comes out of the blowhole, so when the hot air hits the cool air it becomes a spouting cloud of steam.

Different kinds of whales can be told apart from one another by the way they spout. The spout of a sperm whale slants forward and to the left. Right whales have a double, V-shaped spout.

Whales' ancestors also had changes in their ears for life in the water. If you swim,

Cross Section of Fin Whale Earplug

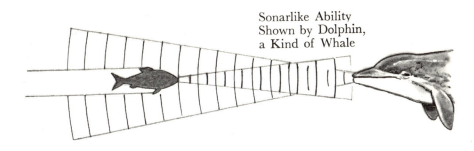

Sonarlike Ability Shown by Dolphin, a Kind of Whale

you know that getting water in your ears is uncomfortable and can cause hearing problems. Some people put wax plugs into their ears when they go swimming, and that's just about what nature did for whales. They have long, cone-shaped "earplugs" made of much the same sort of material as your fingernails. These plugs actually grow as a whale grows, with a new layer of plug forming over an old one every six months or so. Thus, if a plug is sawed in two, it shows "rings," like the rings of a tree trunk. Just as it's possible to tell a tree's age by the number of rings of its trunk, it is sometimes possible to judge a whale's age by the rings of its earplugs.

It might seem that having plugs in their ears would make whales hard of hearing, but this isn't the case at all. Whales can hear very well, and their sense of hearing is of great importance to them because two other senses—smell and sight—can't be counted on. It's almost certain that whales cannot smell at all, and it's easy to understand why: How could an air-breathing animal smell anything underwater when a single sniff would fill its lungs with liquid? As for sight, that's not of much use to an animal that spends a lot of time down in dark water where no sunlight reaches, as some whales do. So, while most whales can see well enough, some really have little need of eyesight, and some are even blind.

But hearing is of importance to all whales, and not only can most of them hear ordinary sounds a good deal better than we can, they are also able to hear many high-pitched and low-pitched sounds humans

cannot hear at all. Whales often make such sounds in communicating with one another.

In addition to superior hearing, some kinds of whales—perhaps all of them—possess a special sense like the sonar equipment used by ships. In using sonar, a ship sends out "ping" sounds by radio. If a sound hits something, it bounces straight back to the ship. Fishing ships use sonar to locate schools of fish, and some whales do exactly the same thing. As they swim, they send a constant stream of hundreds of "clicks" shooting out ahead of them. If the clicks hit something, they come bouncing back, and in some fashion they tell the whale what's in front of it. Thus, even in the darkest waters, these whales can easily locate the creatures they eat and also avoid bumping into rocks and other objects. In darkness, where eyesight would be useless, the whales literally "see" by hearing!

There are two basic kinds of whales—*toothed* whales and *baleen* whales. Toothed whales have teeth, of course, but they don't necessarily have lots of teeth; some have no more than two. The food of toothed whales is usually small fish or squid, but one kind of toothed whale eats such things as penguins and seals, while another sometimes dines on giant squid.

The other kind of whales, baleen whales, have no teeth. Instead, they have rows of thin, triangle-shaped pieces of the material called baleen hanging from each side of their upper jaw. Baleen is also known as *whalebone,* but it really isn't bone at all. It's made of the same sort of stuff as our hair and fingernails. There may be from 100 to

Blades of Right Whale Baleen

Cross Section

Baleen Inside Upper Jaw

400 pieces of baleen on each side of a whale's upper jaw. Each piece is about a third of an inch thick, and there is about a quarter of an inch space between pieces. From the outside, a baleen whale's whalebone looks like fringe hanging inside its mouth. From inside a whale's mouth, the baleen looks like a rough, hairy doormat, because the inside edge of each triangle forms hundreds of hairlike bristles, all tangled among the bristles of the pieces of baleen next to it.

A baleen whale uses this odd mouth machinery to get its food. All baleen whales eat mostly tiny shrimplike creatures, known as *krill,* that live by the millions in various parts of the ocean. A hungry baleen whale swims through a swarm of krill and either gulps up a huge mouthful of krill-filled water or else just lets the water flow through its open mouth. The water either flows out between the pieces of baleen or is squeezed through them by the whale's tongue. But the wriggling krill can't get through the tangled, hairy mat formed by the inside edges of the baleen, so they stay trapped in the whale's mouth and are swallowed.

Most scientists think the ancestors of whales were rather small, ratlike creatures that lived well over 60 million years ago. Some of these creatures probably lived along seacoasts, or perhaps along rivers, and must have mostly eaten fish, which they caught by diving. They were surely fine swimmers. Perhaps they were a bit like the otters of today. During many millions of years, changes in their bodies made them able to hold a lot of air inside themselves,

Krill

pushed their nostrils up to the tops of their heads, turned their front legs into flippers and their tails into flukes, and did away with their back legs. In time they became the creatures we now know as whales.

But even the first kinds of true whales weren't much like the whales of today. The oldest true whales we know of, from fossil skeletons, lived about 50 million years ago. It appears as if they probably spent most of their time in shallow water, which most whales of today don't do. And, from the way their bodies were constructed, it looks as if they may even have been able to crawl up onto land once in a while, which no whale of today can ever do.

One kind of long-ago whale probably looked more like a sea serpent than like a whale. Its 70-foot-long body was snakelike, and its head was much like a lizard's head. It was so much like a snake or lizard that the scientist who found its fossil bones thought it was a reptile and named it Basilosaurus, which means "king lizard." It was only after another scientist more carefully studied the creature's teeth that he realized it was actually a mammal, and a whale. He renamed it Zeuglodon, which refers to the shape of its teeth.

When the first whales appeared, there were very few of the kinds of mammals that are around now. Thus, whales are not only the biggest and strongest of all the world's animals, they're also one of the oldest of all the orders of mammals on the earth today. They certainly are distinguished relatives for the rest of us mammals to look up to!

Possible Whale Relative of 50 Million Years Ago

Plankton (Greatly Enlarged)

The Rorqual Whales—BIGGEST OF THE BIG

IN THE BRIGHT, sunlit water just beneath the surface of the sea that surrounds the great, ice-covered continent of Antarctica, an enormous mass of countless billions of living creatures is spread out like a great cloud.

A large portion of the living things that make up this cloud are plants. But they are *strange* plants! They are jellylike blobs of green encased within transparent, glassy, two-piece shells. They are shaped like circles, squares, and six-pointed stars. And, they are so tiny they can only be seen through a microscope.

There are also many animals in the great cloud, and many of them, too, are tiny and very strange in appearance. But most of the animals are all one kind of creature that is quite large compared with all the others—about 2¼ inches long. They are shrimplike things, with many jointed legs, long feelers, and bulgy eyes, like tiny brown beads, on stalks. Their bodies are orange-red, and there are so many of the creatures—millions of them—that big patches of sea hundreds of yards wide are colored orange by them!

This colossal crowd of living things—plants and animals—is known as *plankton,*

which means "wandering" or "floating." It is often called "the pasture of the sea," for it is really very much like a great meadow on land. The tiny plants, which are called diatoms, are the "grass" of the sea. And the shrimplike creatures, called krill, are really much like cows or sheep, for they graze on the diatoms, the sea grass.

However, just as a sheep might become food for a coyote or wildcat, the shrimplike krill are the food of a great many sea creatures. They are eaten by many kinds of small fish, some kinds of squid, penguins and other seabirds, and the seal known as the crabeater seal. And, tiny though they are, krill and similar creatures that are even smaller are the main food of the biggest animals in the world—one of which is now swimming eagerly toward the great cloud of plankton.

It is an enormous fishlike shape that comes slicing into the orange-tinted water like a giant torpedo. Beside it, the largest of all elephants would seem like a toy; a human would be as insignificant as a flea. It is the king of animals, the largest animal known to have ever lived on the earth—a blue whale.

Its great, curved mouth opens, displaying

Comparative Sizes

Blue Whale, Man, and Elephant

a black fringe of baleen. Water, filled with thousands of wriggling krill, comes flooding into the cavernous throat. The mouth closes, jets of water squirting from each side of it. The red mass of krill is swallowed down. Thus, the largest of all creatures feeds upon one of the smallest.

The blue whale belongs to a family of whales that is known as the rorquals, or fin whales, because they each have a triangle-shaped fin on their back, near the tail. There are seven kinds of rorqual whales, and six of them are giants, but the blue whale is the giant of them all. Many of the largest blue whales caught by whalers were at least 90 feet long and weighed well over a hundred tons each—as much as 1,400 people, each weighing about 150 pounds. The largest one ever caught was 107 feet long and probably weighed about as much as a crowd of some 2,000 average-sized people!

Yet, despite its enormous size and weight, there is nothing clumsy or slow about a blue whale. It is a graceful, streamlined, elegant animal that can move fast. A 90-foot blue whale swimming at normal speed moves about 16 or 17 miles an hour, and a blue whale in a real hurry can keep up a speed of some 23 miles an hour for about ten minutes—which is faster than any human can run.

Although blue whales generally feed just at or a little below the surface of the water, they can go much deeper, and they often will if frightened. These big creatures can dive down as far as one-fourth of a mile and can stay underwater for as long as 40 min-

utes. When a blue whale comes up after a long stay underwater, it blows a pear-shaped cloud of steamy breath some 20 feet straight up into the air.

Blue whales get their name from the grayish-blue color of their back and sides. But they are also known as sulfur-bottom whales, because many of them captured by whaling ships had a sort of yellowish crust on their undersides, which resembled the yellow chemical sulfur. This crust isn't a natural part of the whales, however. It gets "painted" on them as they swim through the huge clouds of diatoms when they feed. Millions of the diatoms, which have a greenish-yellow color, get stuck to the whales' undersides, forming the yellow coating.

There are actually three kinds of blue whales. One kind lives in the oceans south of the equator and never goes north of the equator. It's known as the southern blue whale. Another kind lives north of the equator and never goes south of it, and, of course, this animal is known as the northern blue whale. The southern blue whale is just a little bigger than the northern one, but otherwise they're exactly alike, and they live in exactly the same way. When it's winter in their part of the world, the whales stay fairly close to the equator, where the water is warm. During this time they mate and have their babies. When summer comes, the clouds of plankton in the seas near the North and South poles increase tremendously in size, so the whales swim toward the poles and spend about four months there, fattening up. The southern

"Pleated" Throat
of a Rorqual

whales swim south to the Antarctic Ocean, and the northern whales swim to the Arctic Ocean, near the North Pole, to feed on the krill there, which is slightly different from that at the South Pole. (Of course, when it's winter in one half of the world, it's summer in the other half; so while one kind of whale is mating and having babies, the other kind is at one of the poles feeding.)

The third kind of blue whale is known as the pygmy blue whale. The word *pygmy* means "very small," so you might imagine that a pygmy blue whale is considerably smaller than its two relatives. Actually, it isn't; a full-grown pygmy blue may be as much as 76 feet long! The name really isn't a very good one. However, the pygmy blue is a bit different from the other two blue whales in several ways, including the fact that it's more of a silvery-gray in color. But, like the other blue whales, the pygmy blue is a krill eater.

It has probably occurred to you that such an enormous creature as a blue whale must have to eat a titanic amount of krill in order to keep itself going. And, indeed, it does. For one thing, blue whales and all the other rorquals are apparently able to take in extra-large mouthfuls because they seem to have expandable throats! The underside of a rorqual whale has a great many long grooves that run from the chin down to about a third of the way from the tail.(In fact, the name *rorqual* comes from two Norse words that mean "grooved whale.") These grooves make the whale's throat and belly look as if they are "pleated." No one has ever been able to see it happen, but scientists are sure those pleats open out like an accordian as the whale is feeding, making its throat much wider and able to take in bigger gulps of krill at a time. As for the total amount of krill that a blue whale must eat before its hunger is fully satisfied, that amount comes to no less than 1 ton—2,000 pounds—for a single meal. However, the whale digests its food quite quickly and is ready to eat again in about three or four hours. So it might eat as much as 3 or 4 tons a day—which would be about the same as eating half of a very large elephant!

However, no blue whale eats that much all the time. They do most of their eating and fattening up when they're at the North or South Pole. Apparently, they don't eat much while they're spending the winter in warmer water, because when they head for the poles they're usually rather thin.

Naturally an animal the size of a blue whale is generally safe from all the large, fierce creatures that make the sea their home. Even the great white shark, which will gobble up turtles, seals, dolphins, and other fairly big animals, would never dare tackle a blue whale. But the whale does have an enemy that's willing to attack it, and, surprisingly, this enemy is very tiny compared with the whale—no more than 3 feet long. It's an exceptionally nasty customer known as a lamprey.

A lamprey is a kind of fish that has a snakelike body and no jaws. It has a round mouth that's like a suction cup filled with pointed teeth arranged in circles. It might well be called the "Dracula" of the sea because it's a bloodsucker! It feeds by fasten-

16

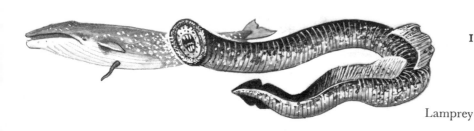

Lamprey

ing its mouth to another creature's body, gouging a hole with its sharp teeth, and swallowing the blood that flows. A lamprey can kill many kinds of fairly large fish by draining them of so much blood that they become weak and sick. It can't do that to a whale, of course, but it apparently attaches itself to blue whales, as well as other whales, and feeds on their blood until something causes it to release its hold. Many blue whales and other whales have been found with scars on their bodies left by the tooth-filled, suction-cup mouth of a lamprey.

However, blue whales have had a far more dangerous enemy than the lamprey—humans. So many of these giants were killed by whalers during the past 90 years that for every hundred blue whales that once swam in the sea, there are now only six!

After the blue whale, the second largest rorqual is the fin whale, also known as the common rorqual. The biggest fin whales are about 80 feet long. These whales are shaped much the same as blue whales, except they're slimmer. They're also colored quite differently—and rather unusually. A fin whale's plates of baleen are all grayish blue except for a cluster of those at the front of the right side of the mouth, which are yellowish white. And the underside of the whale's chin is white on the right side, but gray on the left side, like its back and sides. It's almost as if a giant with a bucket of white paint came along and "touched up" all fin whales in this odd fashion!

Just as there are northern and southern blue whales, there are also northern and southern fin whales. They live in much the same way as blue whales, except that in addition to eating krill, fin whales also feed on several kinds of small fish. And, whereas blue whales are loners, often found by themselves except at mating time, fin whales seem to prefer company. They usually stay together in groups made up of just a few or as many as several hundred fin whales.

The rorqual called a sei whale reaches about 60 feet in length. Its body is bluish black, and its baleen is black, sometimes with streaks of white. The bristles on the inside edges of the baleen are white and form such a thick, shaggy tangle.that it resembles sheep's wool.

The fourth largest rorqual, Bryde's whale, looks so much like a sei whale that it has even fooled experts. However, its baleen is quite different from that of the sei —only half as long and with stiffer bristles on the inside edge. The baleen is also colored differently, being whitish with grayish stripes at the front of the whale's mouth and grayish black toward the rear. Bryde's whale has a slightly different appetite from the other rorquals, for it eats mainly fish, and it has even been seen gulping up 2-foot-long sharks. At the very most, a Bryde's whale may be 50 feet long, but usually they are slightly smaller.

The runt of the rorqual family is the Minke whale, which is usually no more than 30 feet long. It looks like a small, rather chubby fin whale, but it can easily be told from the fin whale and all other rorquals by the color of its baleen, which is all yellowish white. The Minke whale seems to

18

Fin Whale

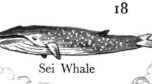

Sei Whale

Comparative Sizes

Bryde's Whale

Minke Whale

Humpback Whale
(Head-on View)

prefer shallow water to the open sea, for it is often found near shore. It eats krill and various kinds of fish, even such large fish as cod and haddock.

There's still another kind of rorqual, but it is really quite different from all the others in a number of ways. The blue, fin, and other rorquals are streamlined and smooth skinned, with short flippers. This creature has a stout and lumpy look, knobby bumps all over its head and snout, and extremely long, rather ragged-looking flippers that also have knobby bumps along their front edge. This rather homely character is known as the humpback whale, not because of its shape, but because of the way it humps its back when it dives.

A full-grown humpback may be 50 feet long and might have flippers as much as 14 feet long (while a 90-foot blue whale has only 7-foot flippers). Seen coming through the water head-on, a humpback looks like a huge, warty ball with a pair of thin, awkward wings flapping at its sides! To add to its knobby appearance, a humpback's head is quite often dotted with barnacles that have fastened themselves to its skin, as they do to the undersides of ships.

Although humpbacks may not be handsome, they are among the most playful of all whales. It's hard to imagine an animal as long and heavy as a railroad boxcar leaping clear out of the water as it turns somersaults, but humpbacks have been seen doing just that! They also roll about, threshing the water with their long flippers while they whistle, squeal, hoot, and make a number of other strange noises. Whale experts point out that there may be some reason other than just playfulness for doing such things, but nevertheless, the whales certainly seem to be doing it just for fun.

Humpback whales also "sing." Not a humanlike song with a melody, of course, but a series of sounds that are repeated, like the song of a bird. But the song of a humpback is much longer than any bird's song, and it is a mournful, eerie, almost frightening thing to hear. First may come a series of long, shrill, echoing wails—lonely sounds, each a little higher than the last. Next, a number of whistling squeaks, slowly dropping lower. Then a strange, low-pitched sound, almost like a human voice resounding through a great cavern—*zoop—zoop—zoop*. The song may go on for as much as half an hour, with the whale's "voice" sometimes dropping to a deep rumble, at times rising like a high note on a violin.

The songs of humpback whales have been recorded by scientists, and it seems as if each whale sings a slightly different song from that of any other whale! Some scientists believe these songs may be the way that whales keep in touch with one another. For it is possible that whales can hear each other's songs even when they are many miles apart from one another.

Like most of the other rorquals, the humpback eats mostly krill and some small fish. And, like the blue and fin whales, there are northern and southern humpbacks that migrate northward and southward in the summertime, swimming back toward the middle of the world in wintertime to mate and have their giant babies.

The Right Whales—PREY OF THE WHALERS

A HUNDRED and fifty or sixty years ago, when anyone said the word *whale,* they were almost always talking about one special kind of whale—the Greenland right whale, also called the bowhead. For these whales of the far northern seas were once so numerous and so much easier to hunt than other kinds of whales that they were almost the only whales hunted and the only whales anyone knew much about. Today, because they were so widely hunted, there are very few of them left—perhaps no more than 2,000 or so.

Greenland right whales are baleen whales and krill-eating whales, like the rorquals. But they are very different from the rorquals in several ways. For one thing, they don't have a fin on their backs, as rorquals do. For another, they don't have grooves on the throat. Their plates of baleen, or whalebone, are much longer than the baleen of rorquals. And they feed in a slightly different way from rorquals.

A rorqual takes a big gulp of krill-filled water, then closes its mouth and squirts the water out through its baleen, leaving all the krill that were in the water caught in the thick, hairy mat formed by the inside edges of the baleen. But a right whale holds its mouth open as it swims through krill-filled water, simply letting the water flow through its baleen until a large number of krill are trapped. Then it closes its mouth and swallows them.

Another difference between a right whale and a rorqual is in their spouts. Rorquals blow a pear-shaped spout out of a single nostril. A right whale has two nostrils, so it spouts *two* jets of steamy breath that form a white, cloudy V.

A full-grown Greenland right whale may be about 65 feet long. Just about one-third of that length is made up of the animal's huge head, the top of which is curved and narrow, like a thick, giant bow—which is how the whale got its other name, bowhead. Its mouth, too, forms a great curve, and from each side of the upper jaw hang as many as 300 triangles of baleen that may be as much as 14 feet long—four times longer than the baleen of even the giant blue whale. A hundred and more years ago, that baleen was worth a great deal of money to whale hunters. It is tough and springy, and a lot of useful things were made from it in those days.

Bowheads aren't graceful and streamlined as most rorquals are. Their bodies

BLACK RIGHT WHALE

swell out like fat bottles, and their flippers are like broad paddles. They are much slower swimmers than rorquals. Their color is usually coal-black, with a single splotch of white, called a "vest," on the chin. Many of them also have "beards" of a few, scattered white hairs on their chins.

Bowhead or Greenland Right Whale

Black Right Whale

These whales, like polar bears, are strictly animals of the far north. They can be found only in the ice-filled waters of the Arctic Ocean. They either stay by themselves most of the time, or in little groups of two or three; although when they move slightly south in the fall, as many as fifty may travel together.

A second member of the right whale family is known as the black right whale. Black right whales, which may reach a length of about 53 feet, seem to like a slightly warmer climate than Greenland right whales do, for they are never found as far north as their bigger relatives. Some of them live in the northern and southern parts of the Atlantic Ocean, and some live in the northern and southern parts of the Pacific. They feed on krill and other tiny shrimplike animals of various kinds.

The black right whale is generally solid black or brownish black and closely resembles the bowhead. However, black right whales have a number of odd, yellowish or pinkish bumps on the top of the head, just in front of the blowhole. The biggest of these bumps is known as the "bonnet." For some reason, barnacles often fasten themselves to the bonnet and other bumps, and this makes many black right whales look as if they have rock gardens on their snouts!

Thus, it's usually easy to tell a black right from a Greenland right.

There's also a third member of the right whale family, the pygmy right whale, which is the smallest of all baleen whales. It reaches only about 20 feet in length. Unlike its two relatives, it has a small, rather bent-looking fin on its back, but otherwise it is much like them. It is blackish, with a white splotch on the underside running from the throat to the belly. Pygmy right whales live in the southern seas around Australia and New Zealand, South Africa, and South America. They are rather rare, and not very much is known about them.

Actually, it is probably lucky for pygmy right whales that they are rare, for there were never enough of them to make it worthwhile for whalers to hunt them. But the other two right whales have been hunted for many hundreds of years, long before most other kinds of whales were hunted. The main reason for this was that right whales are slower swimmers than most kinds of whales, and are therefore easier to catch. And, even more important to whale hunters, right whales don't *sink* when they are killed, as most other whales do; their very thick blubber keeps them afloat. Thus, there was no danger of losing them as there was with other whales. So, whalers said they were the "right" whales to hunt—which is how the whales got their name.

The chase and capture of a right whale 150 years ago would have begun with a keen-eyed sailor peering into the distance from high up near the top of a whaling ship's tallest mast. Slowly, the man turned

Close-up
Black Right Whale

Pygmy Right Whale

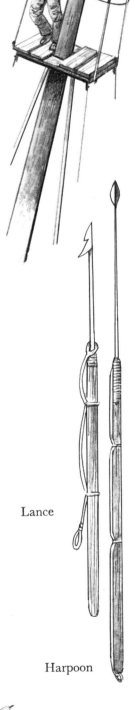

his head from side to side, scanning the vast, gray-green expanse of ocean spread out on all sides.

Abruptly, he saw what he had been looking for. In the distance, a twin spout of white had shot up above the water—the "blow" of a right whale. Taking a quick breath, the sailor yelled out the words that meant a whale had been sighted. "There she blows!"

From the deck far below, an officer's voice yelled back at him. "Where away?"

"Dead off the starboard bow."

At once, the ship broke into a bustle of activity. Men piled into two of the boats that hung along the sides of the ship, and the boats were lowered into the water. There were six men in each boat—four to pull the oars, an officer at the back to steer, and a harpooner at the front. The instant each boat's bottom was in the water, the oarsmen began to row, driving the craft in the direction the lookout had indicated.

Shortly, they could see the black curve of the whale's back and bump of its blowhole above the water. The oarsmen pulled with all their strength. There was no sound save the steady, faint *clumpa-clumpa* of the oars rubbing against the oarlocks. No one spoke, for they knew the whale had keen hearing, and if it were frightened by a noise, it might dive and they would lose it.

They drew closer—closer. Quietly, the harpooner in the nearest boat rose to his feet. He positioned himself with one foot in the bottom of the boat, the knee of his other leg braced against a notched board in the boat's front end. He drew his arm back, carefully balancing the long, heavy harpoon and eyeing his target. Then, with a quick heave, he hurled the weapon into the broad, black back.

At once, the whale dived. Attached to the harpoon was a sturdy rope, many hundreds of feet long, that lay coiled in a tub in the bottom of the boat. The coils were whipping out of the tub almost faster than an eye could see as the whale swam at top speed, trying to escape the sudden, shocking pain it felt. Towed by the rope, the boat rushed through the water, sending out a sheet of spray on each side. This was what New England whalers called "a Nantucket sleigh ride"!

After a time, the whale tired. Once again, it lay at the surface of the water. Rowing slowly and as quietly as they could, the whalers brought their boat up behind the huge animal. What they were about to do was the most dangerous part of whaling, and perhaps at this moment some of them were thinking of the motto of whalers: "A dead whale or a smashed-up boat!" For, now they had to kill the whale, but in trying to do so, they themselves might be killed.

Closer, closer to the whale moved the boat. The harpooner, now holding a long, slender lance, stood ready. The front of the boat bumped the whale's side. Whalers called this moment "wood to black skin." At the instant the boat touched the whale, the harpooner drove the lance deep into the animal's body at a point where he knew the lance would enter its lungs. At once, the oarsmen were desperately pulling as hard and fast as they could to get the boat

Lance

Harpoon

Nantucket Sleigh Ride

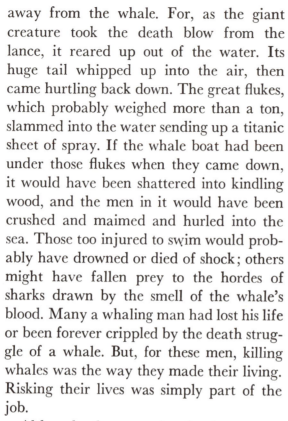

away from the whale. For, as the giant creature took the death blow from the lance, it reared up out of the water. Its huge tail whipped up into the air, then came hurtling back down. The great flukes, which probably weighed more than a ton, slammed into the water sending up a titanic sheet of spray. If the whale boat had been under those flukes when they came down, it would have been shattered into kindling wood, and the men in it would have been crushed and maimed and hurled into the sea. Those too injured to swim would probably have drowned or died of shock; others might have fallen prey to the hordes of sharks drawn by the smell of the whale's blood. Many a whaling man had lost his life or been forever crippled by the death struggle of a whale. But, for these men, killing whales was the way they made their living. Risking their lives was simply part of the job.

Although the men in the boat were drenched by the great splash, the last blow of the whale's tail had missed their boat. They were alive, and the great beast they had sought to kill was dead. They began to tow the huge black body toward their ship, which was moving to meet them.

The dead whale was fastened alongside the ship, and the work of "cutting-in" began. The ship's officers, wearing metal spikes strapped to their boot heels to keep from slipping, climbed onto the big carcass. They carried long-handled spades with knife-sharp edges, and with these tools they sliced into the whale's flesh, cutting away long strips of skin with blubber attached.

A hole was cut into the end of each strip, and a hook was slipped through the hole. Crewmen tugged on the ropes, pulling the strips of skin and blubber off the whale's body and onto the deck of the ship.

Once all the blubber was stripped off the dead whale, and the baleen cut out of its mouth, the big body was unfastened from the ship and simply allowed to float away. Shortly, sharks would find it and begin to tear away chunks of the rich, red meat that clung to the bones. Seabirds would alight on the carcass to peck at the vast source of food. The body might float until it came to rest against a part of the broad expanse of ice that covered much of the Arctic Ocean, and then a wandering polar bear might find it and have a great feast.

Meanwhile, on the deck of the whaler, men went to work cutting the huge strips of blubber into large chunks known as "horse pieces." The harpooners had the job of turning these pieces into oil. The pieces were thrown into two huge pots, set over a crude furnace, in the midst of a wall of brick. The blubber rapidly cooked down into oil which the harpooners ladled out of the pots and poured into a big copper tank to cool. As it cooled, it was poured into wooden barrels.

There was a lot of blubber to be cooked, and the cooking, which was known as "trying out," took a long, long time—often as much as an entire day and night. By the time the hard, tiring work was finished, the men and the deck of the ship were soaked with oil. The oil on the deck was mopped up and added to the oil in the containers,

Cutting Spade

Horse Pieces

then the deck and the men were vigorously washed down with seawater. Finally, the barrels of oil were carried down and stored in the cool darkness of the broad space known as the hold, in the bottom of the ship.

Today, most people think of those old-time whalers with anger! Over the years, the whalers slaughtered so many whales that they nearly wiped out all the Greenland right whales, the black right whales, and several other kinds. It seems dreadful to us, now, that such a thing could have been done.

But, we have to understand that what those whalers were doing was actually very necessary at that time. Whale oil and whalebone were extremely important to people 150 years ago—as important as electricity, gasoline, or plastics are for us today!

Our streets and highways are lit at night by bright electric lights to help people find their way about in safety. But there were no electric lights 150 years ago. The lights that glowed on city streets then were made by burning *whale oil*. And most of the lamps in people's homes burned whale oil, too. It gave a bright light and had no smell. It was the best kind of lighting that people had in those days.

Whale oil was also used to make soap, paint, and a number of other very useful things. Today we can make those things without it—but, 150 years ago it would have been difficult to do so.

The people of 150 years ago didn't have plastics or many of the tough, springy metals that we can now make. The closest thing they had was whalebone. Springs for horse-drawn carriages were made from whalebone. Whalebone was used for making umbrellas, ladies' corsets, suitcases, chairs and sofas, and a good many other useful and necessary things. For anything that needed a tough, springy material, whalebone was just about all there was.

Another valuable material was ambergris, which comes from sperm whales. It is a dark, lumpy, rather smelly stuff that comes from the whale's intestines. Long ago, people discovered that the finest perfumes could be made with ambergris. Today, we can make ambergris artificially, but 150 years ago that wasn't possible. Real ambergris was extremely valuable to perfume makers and they were willing to pay a good deal of money for a chunk of it. So, whalers were always on the lookout for it.

Thus, there were a lot of reasons for hunting whales long ago, and the men who hunted them were doing valuable, important work at that time. It is true that they did some stupid things, such as overhunting whales until many kinds were nearly wiped out. But, we cannot really condemn the whalers of long ago for hunting whales at a time when most of the people of the world *wanted* them to do so.

The people of the past nearly wiped out the right whales. It's up to us, today, to make sure that such a thing never happens again!

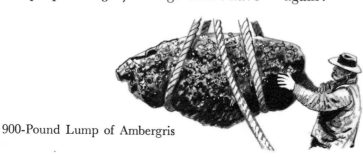

900-Pound Lump of Ambergris

Giant Squid

The Sperm Whale—HUNTER OF THE KRAKEN

IN THE ABSOLUTE darkness, nearly half a mile down beneath the surface of the sea, a huge creature was moving swiftly through the water. To anyone who could have seen it, it would have seemed like a nightmare. A mass of incredibly long, snakelike arms, their undersides covered with big suction disks, streamed out of a torpedo-shaped body. A pair of huge, humanlike eyes stared unblinking into the darkness—a giant squid!

For thousands of years such creatures belonged among humankind's legends of dragons, sea serpents, and similar mythical monsters. Norse seamen of olden times had spoken fearfully of the Kraken—the gigantic octopuslike thing that sometimes rose up out of the depths of the sea to crush whole ships in the horrible embrace of its many writhing arms. Serious-minded people had scoffed at such tales. But, within only the last hundred years, scientists have found that there *are* "krakens"—giant squid that, while not as gigantic as the legendary monster, are nevertheless of awesome size. And this creature, swimming through the darkness, was such an animal. From the tips of its stretched-out arms to its arrowlike tail, it was a good 50 feet long!

Abruptly, the squid slowed its speed. Some sense, some special ability, had suddenly warned it that it was close to danger. In the darkness nearby, there was something that even this huge creature feared. It sought to turn and flee—but too late!

Out of the blackness a darker blackness suddenly loomed; an enormous shape dwarfed the great squid. It was a 60-foot-long male sperm whale, the giant squid's greatest enemy.

The whale's mouth gaped open. Quickly, the squid's powerful arms wound themselves around the whale's head; the suction disks exerted a fierce grip. The squid was desperately trying to keep from being swallowed!

Now, holding its prey in its mouth, the whale went streaking upward, through water that grew ever brighter, until at last the broad back and upper portion of the head broke through the surface into the air. Instantly, a 12-foot-long cloud of white vapor shot diagonally from the whale's head, with a sound like an explosion, as the animal's breath jetted out of its blowhole. Then, after taking as many as 50 breaths, the whale began to chew up the squid into large chunks and swallow them.

Of course, no one has ever actually seen a sperm whale capture a giant squid down in the sea's black depths. But there is no doubt that struggles between these two giants do take place; there is evidence for them. Many a sperm whale has been seen with rows of round, 4-inch-wide scars on its head and jaw made by the hard edges of a squid's suction disks. In the days of whale hunting, when thousands of sperm whales were killed and cut up, many were found to have the remains of giant squids in their stomachs. And, the best evidence of all, a portion of such a struggle that was somehow carried up to the ocean's surface was seen. The whale appeared to have the squid's tail in its mouth, but whether it finally swallowed its prey, or whether the squid managed to get away, isn't known.

Actually, such fights are probably not very common. For one thing, it seems as if it is mainly only the big male sperm whales that will even attack giant squid. Few female sperm whales, which are much smaller than the males, have ever been found with either battle scars or with the remains of giant squid in their stomachs. And, for the most part, both the male and female whales usually satisfy their hunger by eating large amounts of ordinary-size squid, only 1 or 2 feet long.

To even go where giant squid live, a sperm whale has to go farther down into the ocean depths than any other kind of whale can go. It has been found that sperm whales can dive down more than half a mile, which is far, far deeper than a human in any kind of diving suit can go. That far down, the water is totally black because light cannot penetrate through it to such a distance.

It may seem puzzling as to just how a sperm whale is able to find squid, giant or otherwise, in that utter darkness. It certainly can't *see* them. The answer is that sperm whales possess a kind of sonar, such as submarines use for detecting things in the water around them. The whales make constant clicking sounds that travel rapidly through the water ahead of them. When these sound waves hit something, such as a squid or fish, they bounce straight back at the whale. In other words, they *echo*. And, somehow, from that echo the whale can tell how far, and in what direction, the thing is that the echo comes from. Thus, even in the total darkness half a mile down, a sperm whale can locate other creatures and can tell which way they are moving.

It might seem that the whales could do just as well by staying up near the surface and hunting creatures they could see. They often do this, but near the surface they have to share the available food with many other creatures. By going way down, where no other large sea mammal, sea reptile, or seabird can go, the sperm whale has a lot more food available to it.

The sperm whale is the biggest of all the toothed whales. A male sperm whale that's full grown may be as much as 60 feet long, just about as long as the big, long-necked dinosaur, Brontosaurus. However, in the 1800s, one whaling ship captured a sperm whale that was 76 feet long; while another ship claimed a record-breaking whale of 84 feet in length!

SPERM WHALE

Giant Squid

Spermaceti

Skull

A sperm whale's head is much bigger and differently shaped from the heads of other kinds of whales. It makes up one-fourth of the sperm whale's whole length and one-third of its entire weight. The reason for this is the sperm whale's head contains an enormous lump of muscle and fatty tissue that weighs about 15 tons, and that is filled with about 4 tons, altogether, of a clear, oily liquid. This liquid is called *spermaceti,* which means something like "seed of the great sea animal," and it is from that word that the sperm whale gets its name. The big lump that holds the liquid is called the *spermaceti organ.* The main reason sperm whales were hunted by whalers was to get the liquid in the spermaceti organ. When that liquid is cooled it becomes a solid wax and can be made into candles and various other things.

Scientists have never been sure just what the spermaceti organ is for. There have been a lot of ideas about it. One idea is that maybe the organ has something to do with the whale's sonar system. Another idea is that perhaps this organ enables a sperm whale to lie motionless underwater after it makes a deep dive. For when a sperm whale dives, it usually stays underwater for a long time—often as much as an hour or even an hour and a half. During such a long period, it seems as if the whale might travel miles underwater, but apparently it doesn't; it usually comes up very near where it went down. Thus, it appears that the whale may go straight down and then stay there without moving until it comes back up. But it could only do this if it had some way of

changing the density of its body to make it the same as the density of the water around it. Otherwise, the whale would just float right back up to the surface again. So, some scientists think the spermaceti organ may somehow be able to change the whale's density.

Why would a sperm whale want to lie motionless far below the surface? Well, perhaps that's how it catches its food. Maybe it uses its sonar to locate some squid nearby, and then it lies in wait, motionless and silent, until one of them comes near enough to be caught with a sudden lunge.

Although the sperm whale is a toothed whale, it actually has teeth only in its lower jaw. There are teeth in the upper jaw, but they're tiny and don't even stick down through the gum. However, the teeth in the long, narrow lower jaw of a full-grown sperm whale are about 8 inches long, and there are from about 9 to 15 of them on each side. When the whale closes its mouth, the teeth fit neatly into holes in the upper jaw.

Even with only half a set of teeth, a sperm whale is able to chew, but most of the time it simply swallows its food whole —even when it eats such things as seals and 10-foot sharks (and one angry sperm whale once swallowed a sailor from a whaling boat that was chasing it!). Actually, a sperm whale doesn't even use its teeth to catch the things it eats. Young sperm whales don't have any teeth until they're about three years old, but they manage to catch plenty of squid without them. So, apart from chewing up giant squid, what does a

sperm whale use its teeth for? For one thing, it uses them for fighting other sperm whales!

Sperm whales generally stay in herds made up of 20 to 50 female whales and their young and one full-grown male who is the leader of the herd. And, during the mating season, male sperm whales do battle with one another to gain control of as many females as they can. That's when they use their teeth!

Such fights have been seen, and what a sight they were! Two big males will charge straight at each other in attack. They bite and tear huge pieces of flesh from each other's heads. Each whale seems to try to get hold of the other's lower jaw, and they twist and wrestle until the water around them seems to boil! They'll break apart, then charge each other again, over and over, until one finally turns and swims slowly away. The other will simply let it go. Usually, both whales seem to be hurt. Some whalers who once watched such a fight lowered a boat and captured one of the whales that had been fighting. They found that its lower jaw was broken and hanging loose, many of its teeth were broken, and there were bloody gashes all over its head!

The herds of sperm whales usually stay in the warmest parts of the ocean. In summer they are usually found well away from the equator, but in wintertime, when it begins to get cold where they are, they migrate toward the equator, where it is always warm. However, each year, many of the big males leave the warm water and make long, long journeys north or south, all the way to the cold waters near the North and South poles. Scientists aren't sure why these males do this. Perhaps they are the ones who have lost battles for the control of a herd.

Moby Dick

Although most sperm whales have black backs and grayish undersides, an albino, or pure white whale, is sometimes born. Probably the most famous of all whales, the whale known as Moby Dick in the book by Herman Melville, is an all-white, male sperm whale. And, as a matter of fact, Moby Dick was a real whale that Melville heard of and put into his book.

There are two other members of the sperm whale family. One is known as the pygmy sperm whale. It certainly is a pygmy compared with its huge relative, for a male pygmy sperm whale is only about 13 feet long, and a female is just 9 or 10 feet. However, the pygmy sperm whale isn't merely a small copy of the sperm whale; it has a much different shape, more like that of a dolphin. But it, too, has a spermaceti organ and teeth showing only in its lower jaw. These little whales, which are black on top and grayish underneath, live in the Atlantic, Pacific, and Indian oceans. Pygmy sperm whales, too, live mostly on squid.

The other member of the family is called the dwarf sperm whale. It lives in the warm waters of the tropics. As its name suggests, it, too, is much smaller than the sperm whale, being only about 9 feet long. It looks much like a pygmy sperm whale, except it has a taller back fin. It probably also eats squid and fish, but not very much is known about it.

31

Pygmy Sperm Whale

The Gray Whales—WAYFARERS OF THE PACIFIC

A BOATLOAD of people had just met a giant! The people were excited, awed, and perhaps just a bit frightened. The giant was cautious, but it seemed curious—and friendly. It lay quietly in the water alongside the boat and let the people touch its huge body.

The giant was a gray whale, one of the big herd of gray whales that comes, every winter, to the quiet lagoons of Baja California —a finger of land that runs alongside the west coast of Mexico. Here, up until the 1930s, these gentle giants were frequently attacked, year after year, by scores of whaleboats and slaughtered until they were nearly wiped out. But today they are protected. The only boats allowed in the lagoons contain people who come in peace for the thrill of meeting a real live giant.

The gray whale herd is about 11,000 strong. It comes to Baja California beginning in late December and stays there until about April. In these warm waters some of the whales mate, and baby whales, from matings of the year before, are born. In these waters, the baby whales learn their way of life.

In April, or perhaps March, a great journey begins. At first, only a few whales at a time leave, most of them males. Then, dozens and dozens, including the mothers with their babies swimming beside them, start out. Soon, a steady stream of whales is moving northward. Swimming without pause, day after day, they stay in shallow water, seldom more than a mile offshore. It is a good 6,000 mile journey they make, and, in about mid-June, the first of them reach their destination—the waters of the Arctic Ocean that wash against the great mass of pack ice that spreads out from the North Pole.

The whales' reason for going to these waters is to eat, and that's about all they do from June until October! Then, before the bitter Arctic winter begins, and the pack ice starts to spread, the whales head back south.

In the past, their return trip was a time of dreadful danger, because whale hunters by the scores were lying in wait for them. As they neared Alaska they were attacked by parties of Eskimos in canoes. Farther south, they ran into boatloads of Tlingit, Haida, Kwakiutl, and other northwest coast Indians, who were skillful whale killers. Then, from the coast of Oregon down to Baja California, American and European whaling ships were waiting!

32

Tlingit Whaler

Today, things are different. People still "lie in wait" to watch the return of the gray whales, but now they simply want to *see* the big creatures. On clear days during the months of November, December, and January, people by the hundreds gather at high places along the western coast and are thrilled by the sight of a distant gray back sliding through the water, the white puffs of a blow, and the flash of flukes lifting into the air.

For a long time, the old whalers didn't know about the lagoons of Baja California where the whales headed each winter. Then, in 1854, the captain of a whaling ship discovered the place and realized that the whole gray whale herd was at his mercy. His whaleboats could go among them and kill all the whales they wanted!

The captain managed to keep this a secret for a while, so as to have all the whales to himself. But finally, by spying on him, other whaler captains also found the lagoon. Then, whaling ships by the dozens descended on the area, and a terrible slaughter of gray whales began. The whalers killed the big animals apparently without giving the slightest thought to the possibility they might wipe out all the creatures. It was a perfect example of the stupidity and lack of concern of greedy human beings. And, by the early 1900s, no more gray whales could be found. It appeared as if they had been wiped out.

Fortunately, that hadn't happened. In the 1920s, gray whales began to be seen again, although far fewer in number than they had once been. And, despite the fact that the whales *had* nearly been wiped out, and were so few in number they could easily *be* wiped out, the whalers went after them again. And again, the gray whales were nearly wiped out! By 1937 it was thought there were probably no more than a hundred of them left.

However, at that point, due to the pleas of scientists and other concerned people, the United States, Russian, and Japanese governments put a stop to the killing of gray whales in American waters. Whalers could no longer lie in wait for the grays as they came down the coast and were also forbidden to go into the lagoons where the whales spent the winter.

But, most scientists were afraid this protection had come too late. They didn't think there were enough gray whales left to keep the whole gray whale race going. Yet, it did keep going! From about only a hundred members in 1937, it was up to several thousand in the 1950s, and today it's up to 11,000 or more. The gray whales are safe!

Like the rorqual and right whales, grays are baleen whales. A gray whale may have from 138 to 174 plates of yellowish-white baleen hanging from each side of its upper jaw. However, whereas rorquals and rights feed on the swarms of krill or other tiny shrimplike animals that drift near the surface of deep water, gray whales feed in shallow water on creatures that live in or on the muddy or sandy sea bottom. In the far north, grays eat mostly tiny crabs. In the south, in the Baja California lagoons, they seem to eat clams, crabs, and even some fish, such as sardines. A gray eats in much the same way as a rorqual; it takes a big gulp of water and mud, then squirts the muddy water out through its baleen and swallows the creatures that have been left trapped in the hairy mat inside its mouth.

Scientists call gray whales *primitive,* because they are probably much like whales were many millions of years ago. They live in shallow water, as primitive whales apparently did, and they have a good deal more hair than other modern whales do— a throwback to a time when whales were still hairy. Also, they have much bigger hipbones and back leg bones than any other modern whales do.

In fact, gray whales are probably a lot like the primitive whale that was the ancestor of both the rorqual and right whales, because grays are a kind of combination of both rorquals and rights. Like the right whales, grays have no fin on their back; but, like the rorquals, grays have "pleats" on their throats. However, whereas the rorquals have from 40 to 100 pleats that run halfway down the body, gray whales never have more than 4 pleats that are only about 5 feet long. This, too, is a primitive feature.

Gray whales can reach a length of about 45 feet. While some of them are gray in color, many are actually black with whitish blotches all over their bodies, making them look gray from a distance. These whales also often appear to have shaggy patches of moss on their backs and sides. Actually, those patches are clusters of barnacles and of creatures known as whale lice, which resemble shrimp, both of which attach themselves in enormous numbers to gray. whales. .

Gray whales have also been known as Devilfish, which seems an odd name for such gentle animals. The name was given to them by old-time whalers, because of the way mother gray whales would "fight like devils" to protect their babies. The whalers deliberately harpooned baby whales so that the mothers would rush to attack the whaleboat, thus bringing themselves close enough to be killed. (The whalers didn't really want the babies, they wanted the much bigger adult.) The fact that a mother whale sometimes smashed up a whaleboat and injured men in the defense of her baby seemed to annoy the whalers, so they unfairly gave her the nickname of Devilfish!

However, people of today who have been lucky enough to see these big, gentle animals up close certainly wouldn't agree with that name. Although a mother gray whale *would,* no doubt, "fight like a devil" to protect her baby, we now know that if gray whales are approached with friendship, they are creatures of peace that seem quite willing to be friends with us two-legged land animals that were once their worst enemies.

Dolphins and Porpoises—THE LITTLE WHALES

THE BLUE WATER of a broad bay sparkled under a sunny sky. Hardly a ripple disturbed the calm surface. Then, suddenly, there was a swirl and a rush! A quartet of big fish-shaped bodies shot up out of the water, each shedding a gleaming shower of droplets. A dozen and more feet into the air their leaps carried them, wetly shining bodies curving gracefully. Then, with a salvo of noisy splashes they plunged back into the water and raced away, like a group of children engaged in a game of tag. Dolphins at play!

Dolphins, which are actually just a kind of small, toothed whale, certainly seem to be one of the most playful of all creatures. They will often swim in rapid zigzags after one another, launch themselves out of the water in great curving leaps, or shoot straight up and then let themselves fall sideways with a tremendous splash, like a child doing a belly flop into a swimming pool. If a group of tame dolphins in an aquarium finds something floating in the water, they will often use it as a toy. One dolphin will race off with the object balanced on flippers or tail while the others dash in pursuit, each trying to capture the toy for itself. When one dolphin succeeds, the others will then

chase it, and such a game might go on for hours!

People often call dolphins "porpoises," but dolphins and porpoises are actually two different kinds of creatures. The main difference between them is that while many kinds of dolphins have jaws that form a sort of beak, porpoises never have a beak. Porpoises are also usually shorter and stouter than most dolphins. However, porpoises and dolphins, together with pilot whales and killer whales, all belong to one big family.

There are more than 30 kinds of dolphins. The one best known to most people is the bottle-nosed dolphin. That's the one you most often see in aquariums, zoos, and marinelands where groups of them are sometimes trained to put on shows—ringing bells, playing with balls, doing acrobatic stunts, and giving exhibitions of "tail dancing." A bottle-nosed dolphin has a dark gray back, whitish underside, and a rather short, stout beak. A full-grown adult may be 11 or 12 feet long.

Wild bottle-nosed dolphins live in large herds, or schools, throughout the Atlantic and Pacific oceans. They are fish eaters that often hunt by surrounding a school of fish,

Porpoise

Dolphin

Bottle-nosed Dolphin

36

COMMON DOLPHINS

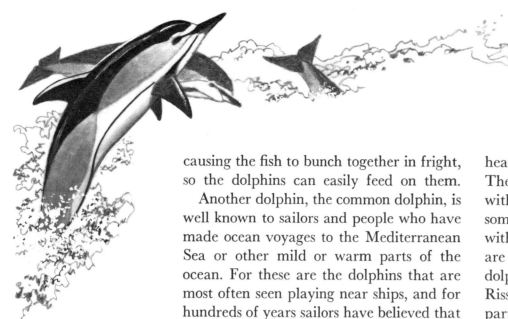

causing the fish to bunch together in fright, so the dolphins can easily feed on them.

Another dolphin, the common dolphin, is well known to sailors and people who have made ocean voyages to the Mediterranean Sea or other mild or warm parts of the ocean. For these are the dolphins that are most often seen playing near ships, and for hundreds of years sailors have believed that the sight of these animals—zigzagging in front of a ship, leaping out of the water, and dancing on their tails—was a sign of good luck for a voyage.

A common dolphin has a decorated look; it has a black back, white underside, and curving streaks of grayish brown, black, and yellow on its sides. Around the eyes is a black strip outlined in white that makes the dolphin look as if it's wearing a mask. A common dolphin may measure around 8 feet long. They, too, live in large schools and eat fish and squid.

The dolphin called a white-sided dolphin, which lives in the North Atlantic Ocean, also looks as if it had been decorated by someone with several different buckets of paint. Its back, tail, and flippers are black, while from its chin to its belly it is yellowish brown with a long, curved, yellowish-brown strip on each side. It has a very close relative, the white-beaked dolphin, that also lives in the North Atlantic and is black on top and white on the underside, including the underside of the beak. Both of these dolphins are about 9 feet long.

The dolphin known as Risso's dolphin looks as if it bumped so hard into something that its nose got pushed in. The front of its head is nearly square, with no beak at all. The body of a Risso's dolphin is dark gray with a whitish underside and looks as if someone had scribbled on it. It is covered with lines, crisscrosses, and squiggles that are probably made by the teeth of other dolphins or perhaps by the beaks of squid. Risso's dolphins, which are found in all parts of the ocean, may be 12 or 13 feet long. They have teeth only in the lower jaw, and they eat mostly squid and cuttlefish.

A certain Risso's dolphin became one of the most famous of all whales. It lived in what is called Pelorus Sound, a portion of water between the North and South islands of New Zealand, in the Pacific Ocean. The dolphin seemed to take great delight in playing around the ships that sailed through the area and almost seemed to be escorting them on their way. The dolphin became such a familiar sight to sailors that they nicknamed it Pelorus Jack. In time, Pelorus Jack became so popular that the government of New Zealand passed a law giving it special protection.

There are a number of other dolphins that have special looks or ways of life that set them apart from others, too. The Chinese white dolphin is milk-white with reddish fins and black eyes. The spinner dolphin gets its name because it has a habit of leaping out of the water and spinning its body round and round, like a top. The rough-toothed dolphin, as its name suggests, has teeth that are wrinkly rather than smooth like the teeth of most dolphins.

The creatures called pilot whales, also known as blackfish, are actually large dol-

Pacific White-sided Dolphin

Risso's Dolphin

Spinner Dolphin

phins that grow as much as 20 feet long. They have black bodies, sometimes with white undersides, and round, domed foreheads that bulge out above their beaks. Many kinds of dolphins, and some whales, have this kind of bulge, which is called a "melon." Pilot whales live in herds of from 50 to 200 or more in the mild and warm parts of the ocean.

There are six kinds of porpoises altogether. The most well known is called the common porpoise, or harbor porpoise. It is one of the smallest of all whales—a chunky creature, only about 5 feet long—with a dark brown or gray back, grayish-brown sides, and a white underside. Common porpoises prefer rather cold, shallow water and are found in bays, harbors, and river mouths. They usually stay in little groups of five or ten.

The spectacled porpoise, which lives in the southern Atlantic Ocean, has a black back and white underside that curves up around its eyes like spectacles. Dall's porpoise, which stays near northern coasts, is black with a broad white stripe slantwise around its middle. The black finless porpoise, which likes warm water, is all black. Porpoises are slow, quiet swimmers, not at all as playful as dolphins. They all eat mostly fish, although some feed on crabs and squid as well.

Dolphins, especially bottle-nosed dolphins, always look as if they are happily smiling, and this makes people feel that the dolphins are both smart and friendly. But, a dolphin isn't really smiling, of course, any more than a crocodile is; its mouth is simply shaped that way. However, it's true that bottle-nosed dolphins, and many other kinds, really do seem to be friendly. A bottle-nosed dolphin that became known as Opo would come into shallow water near a New Zealand beach and allow children to play with it and even ride on its back! Another bottle-nose, nicknamed Georgy Girl, stayed near a Florida beach and also became friendly with children. Most kinds of dolphins seem to be tamed very easily, often just by being touched and talked to; and once they are tame they love to be petted, scratched, and played with by humans. Tame dolphins also seem to enjoy teasing humans and playing tricks on them —such as suddenly splashing people with water to make them jump! For some reason, they seem to like us.

And, dolphins *are* smart. They can learn to do things a lot more quickly than most other kinds of animals can, and they seem able to remember things well. A tame dolphin named Paddy was taught by scientists to pick a certain shape out of a large number of other shapes. It was a rather hard problem, but Paddy had no trouble with it. And seven months later, when he was given the problem again, Paddy showed that he remembered it perfectly.

It even seems, at times, as if dolphins can actually figure out the *reasons* for things, as humans can. Three dolphins at a marine-land in Florida had been trained to jump out of the water, all at the same time, for a show. Their reward for doing this trick was a piece of fish for each of them. During one show, only two of the dolphins made

the jump properly; the third dolphin was late. Because of him, none of the three got any fish. The trainer signaled the three to try the trick again; again the same dolphin was late, and no fish pieces were handed out. But this time, one of the dolphins that had done the trick correctly swam to the late one and bit it on the head. There seems no doubt that the angry dolphin had figured out that it wasn't being given any fish because its companion was "goofing off"—so it punished the "goof-off"! The punishment worked, too, because next time the trick was tried, the dolphin that had been bitten made the jump on time!

Dolphins are often noisy. With their blowholes they squeak, quack, squawk, creak, and whistle. For a time, some scientists believed these sounds might actually be a kind of language and had hopes that dolphins and humans could learn to talk to each other. But it now appears that dolphins aren't *that* smart. Although some of the sounds they make apparently do have a meaning, the sounds aren't truly a language, any more than the sounds made by a dog are.

However, dolphins do communicate with each other. Bottle-nosed dolphins make a special sound that seems to mean "look this way," and also make sounds that show they're excited or scared. Each bottle-nose also has a special "identity" sound, almost a sort of name, and when that sound is used in a certain way, it means the dolphin is in trouble. When an injured female or young dolphin makes such a sound, the other dol-phins of its group will come to help it. If the injured dolphin can't swim, or is in danger of drowning, other dolphins will actually carry it up to the surface so that its blowhole is above water and it can breathe. Often, even different kinds of dolphins will help one another this way. (On the other hand, porpoises swim away from another porpoise that gives a distress sound.)

Dolphins also have a sonar system, as sperm whales and some other whales do. They shoot clicking sounds out ahead of themselves—as many as several hundred clicks a second—and the echoes that come bouncing back tell the dolphins if something is in front of them. Dolphin sonar is far better than the kind of sonar humans can produce with machines. As an experiment, a tame dolphin was once blindfolded and half a vitamin pill was dropped into the water. By means of its sonar, the dolphin located the tiny object at once.

Quite a few scientists in several parts of the world are studying dolphins to try to find out more about them. And dolphins of all kinds will probably continue to be star performers at zoos and marinelands all over the world. Fortunately, dolphins aren't being hunted for any reason at this time, although some of them are sometimes killed by commercial fishermen when dolphins eat fish the fishermen want to catch. Most people don't want to see harm done to these playful, smiling animals. Hopefully, these smart little whales that seem to want to be friends with humans will continue to be treated as friends by more and more people.

Weddell Seal

KILLER WHALES

The Killer Whale—FLESH-EATING DOLPHIN

A WEDDELL SEAL was in trouble. It had dived deep down into the water of the Antarctic Ocean in search of the fish that were its food. When it came back up to the surface, it was suddenly aware of several large, sleek, black-and-white bodies closing rapidly toward it—killer whales!

The seal had only two ways of escaping. It could dive again—for it was able to dive deeper and stay underwater longer than its enemies could—or it could make for the safety of a nearby chunk of ice. The hunters could not follow it there, for their fish-shaped bodies were not able to leave the water and climb up onto ice or land as the seal could. The seal's instincts led it to choose the ice. Exerting every bit of effort it could collect, it streaked through the water, scrambled up onto the white, gleaming mass and crouched there, panting.

But the seal was far from safe. The whales were hungry, and the seal's 8-foot-long, 700-pound body meant food for them, just as a large fish was food for it. And killer whales are smart animals and cunning hunters.

Thud! The tiny island of ice shivered as one of the whales rammed nose first into it at top speed. A chunk of the ice splintered

loose and floated away. Thud! A second whale smacked against the seal's refuge. The seal crouched back in terror as a whale lunged up onto the ice for a moment, its big mouth open and its sharp teeth bared, before it slipped back into the water. Splat! Another whale, with a slap of one of its flippers, sent a shower of water raining down on the seal.

These actions had a purpose. By ramming the chunk of ice with their noses, the whales would soon break it into bits, dumping the seal back into the water. By lunging at the seal, and by showering it with water, they might cause it to panic.

That was what happened. Terrorized, the seal sought only to get away, to flee from the chunk of ice that suddenly seemed to have become a trap! It launched itself into the water and swam desperately for its life. But killer whales are the fastest of all whales and, in fact, one of the fastest of all sea animals. Within seconds, the three big hunters had caught up to their prey.

Killer whales are the only kind of whale that eats warm-blooded creatures—such as seals, dolphins, and penguins—rather than just cold-blooded ones, like fish and squid. These whales are actually giant dolphins;

they are the biggest, fastest, most powerful members of the dolphin family. A full-grown male killer whale may be as much as 30 feet long, weigh nearly 20,000 pounds, and be able to swim a good 30 miles an hour at top speed.

Killer whales are found in all parts of the ocean, but most often in the colder parts, near the North and South poles. They are rather potbellied animals, with shiny black backs and white undersides. Often they have an oval-shaped white spot behind each eye, and they sometimes have a curved splotch of white farther back on each side. Males have extremely long flippers and a tall, pointed back fin that may be as much as 6 feet high. Females are usually only half the size of males and have smaller flippers and smaller, curved fins.

Killer whales swim in packs of from several to as many as 40 animals, both males and females, with the females and any young ones in the center and the males surrounding them. They usually stay close to the surface, with their tail fins sticking up out of the water. The sight of these black triangles slicing through the sea will send seals and penguins scrambling onto shore or a handy chunk of ice!

A pack of killer whales seems to hunt in much the same way wolf packs do on land —using intelligence and cooperation. Nearing one of the thick ledges of ice that coat the water near the North and South poles, the whales will often poke their heads up out of the water to peer across the ice as if searching for signs of life. If there would be some seals or other creatures in sight, the whole pack of killer whales may suddenly dive down into the water. Moments later, with great booms and cracking sounds, the ice will begin to heave and break up as the killer whales, swimming up beneath it, slam their heads into the bottom of it. This breaks the ice into chunks and tilts the chunks sideways so that any creatures on the ice may find themselves suddenly deposited in the water—at the killer whales' mercy! Killer whales have been seen breaking up such shelves of ice that were 2½ feet thick.

A pack of killer whales was once watched as it attacked a school of small dolphins. The whales swam around the dolphins in an ever-narrowing circle, causing them to bunch together in terror. Then, a single killer whale darted among the dolphins and began to feed while the other whales continued to circle, keeping the dolphins huddled together. When the first whale finished its meal, it joined the circle again, and another whale took its place, feeding on the helpless dolphins. One after another, each killer whale ate its fill while the other whales watchfully circled, making sure that no dolphins escaped.

Packs of killer whales have also been seen attacking even big rorquals and gray whales, as wolves will attack a moose. Because of their ferocity in attacking creatures large and small, killer whales have long been looked on as villains—the most savage and terrifying animals in the entire world, ready to attack and eat anything, including people! They were regarded as bloodthirsty beasts that often killed for the fun of it!

False Killer Whale

However, the killer whale's bad reputation is unjustified. Careful studies have shown that while killer whales will eat warm-blooded animals such as seals, penguins, and dolphins, and may well sometimes attack larger creatures, they eat mostly fish and squid, just as other toothed whales do. And, as for eating humans, that has never happened. There have been many cases of killer whales that had the opportunity to attack a human and didn't.

As a matter of fact, killer whales seem to be as curious about humans, and almost as anxious to receive affection from humans, as most other dolphins are. A number of them have been captured in recent years, and they were very quickly tamed and taught to do tricks. They will take food straight from the hands of their trainers and other people, allow humans to ride on their backs, and put on spectacular acrobatic performances for human audiences. And people who have trained both killer whales and other kinds of dolphins insist that the killer whales are much more intelligent. They certainly don't seem to be "bloodthirsty savages."

The killer whale has a close relative that is known as the false killer whale, apparently because those who named it felt that the two animals looked alike. The false killer isn't really much like its relative, however. Its body is completely black and rather differently shaped. It is smaller than a killer whale, never more than 18 feet long. Its habits are different, too, for it eats squid and large fish such as cod, but never any warm-blooded creatures. And, unlike the killer whale, it stays away from the cold polar regions and from the coasts. False killer whales live mainly on the deep high seas, in herds of as many as a hundred or more. Apparently, when these whales get too close to land they get into trouble, for large numbers of them are often found stranded on shores.

Another close relative of the killer whale is the pygmy killer whale, which is only 8 or 9 feet long at most. Pygmy killer whales are slender bodied, with dark gray backs, whitish bellies, and often with a splotch of white on the chin. Sometimes they have whitish lips. They are found in warm tropical waters.

Weddell Seal

44

River Dolphins—FRESHWATER WHALES

WE'RE SO USED to thinking of whales as creatures of the great open sea that many people are surprised to learn there are whales that live in rivers. They belong to a family of toothed whales that are usually called river dolphins.

There are four kinds of river dolphins. Most of them look rather different from one another, but they all have very long, slim jaws filled with little, needle-sharp teeth. And they all have tiny, almost useless, eyes. Some of them can barely see, and one kind is almost blind. But, actually, eyes wouldn't be of much use to most of them anyway, as they all live in muddy, quite dark water. However, they have no trouble at all finding food or keeping from bumping into things, for they have an extremely good sonar, or echo location system.

One kind of river dolphin lives in rivers in India and Pakistan. Called the Ganges River dolphin (or sometimes Indus River dolphin, depending on the river it's in), it has a dull black body and large flippers shaped like fans. The largest of this kind of animal is about 8 feet long.

These river dolphins are almost constantly swimming. They stop only once in a while, for just a few seconds, to take a nap,

which is the only sleeping they do. They go up to the surface of the river, to breathe, about every 30 or 40 seconds. As they swim near the muddy river bottom, they move their heads up and down, apparently using their sonar to locate food. They feed by using the tips of their long jaws to push fish and freshwater shrimp out of hiding places in the mud, snapping them up and swallowing them whole.

Another kind of river dolphin, the Chinese river dolphin, actually lives in a lake —Lake Tungting, in China. It is about 7 or 8 feet long, with a bluish-gray back and white underside. It, too, feeds by rooting in the bottom mud, mostly for catfish.

Amazon dolphins, which live in a number of rivers in South America, are about 7 feet long, with dark backs and pinkish undersides. They, too, get much of their food by grubbing in mud, but they also catch many fish by snatching them out of the water as they swim. And one fish the Amazon dolphins don't hesitate to eat is none other than the savage little flesh-eating piranha, with teeth like razor blades.

The fourth kind of river dolphin lives in a large bay on the coast of South America where two rivers flow into the sea. Called

Chinese River Dolphin

46

Piranha

GANGES RIVER DOLPHIN

the La Plata dolphin, it is the smallest of the river dolphins, only 5 feet long. It is usually pale brown in color.

Unlike other whales and dolphins, which have fishlike shapes with heads that merge right into their bodies, river dolphins all have noticeable heads and necks—almost primitive features that suggest a relationship to the first kinds of dolphins.

It also seems strange that while most whales and dolphins are sea animals, this one little group of dolphins should be freshwater creatures. But most scientists think the ancestors of river dolphins were actually sea animals, too. Perhaps these ancestors gradually moved closer to land, finding a good source of food in the kinds of fish that frequented the places where rivers flowed into the sea. And gradually, perhaps, the dolphins moved into the rivers where there was plenty of food and no enemies such as sharks or killer whales. In time, their descendants would have become the freshwater river dolphins of today.

Beluga or
White Whale

Narwhal

Narwhal and Beluga—TWO UNUSUAL WHALES

THERE IS ONE family of toothed whales that contains only two rather unusual creatures. One member is a whale with a horn on its nose and the other is a whistling, white whale!

The whale with the horn is called a narwhal. It is a rather fat-bodied, round-headed whale that may be 15 or 16 feet long, not counting the 8- or 9-foot-long twisted, pointed horn that sticks straight out from the front of its head. A narwhal is bluish-gray or brownish and is often covered with spots, like a leopard. It doesn't have a back fin, just a short row of little bumps, about 2 inches high, on its back.

These whales are animals of the far, cold North. They live in little groups of from six to ten in the waters of the Arctic Ocean that surrounds the North Pole. They are fast swimmers and deep divers that eat squid, fish, crabs, and shrimp that they grab with their mouths and swallow down whole.

A narwhal's horn isn't really a horn—it's a tooth! Narwhals have only two teeth, both of which grow straight forward, rather than downward, in their upper jaw. One tooth usually never even comes out of the gum. But the other tooth grows and grows and grows, out through an opening in the narwhal's lip, becoming a long, pointed tusk. As it grows, it twists, or spirals, so that when it reaches full length it is like a long, slim screw made of ivory. It is usually only male narwhals that have this tusk; only rarely does a female grow one. It is almost always the narwhal's left tooth that becomes a tusk, although once in a while a right tooth grows instead. And, once in a very great while, *both* of a male's teeth will grow out so that he has a pair of horns sticking side by side out of his nose. Rather oddly, no matter whether it is a left tooth, a right tooth, or both teeth, the spiral of the tusk is almost always to the left.

Every part of a living creature's body usually has some purpose, so it seems there should be a reason why male narwhals have tusks—but no one knows for sure what the reason is. Does the narwhal have its tusk so that it can more easily catch the fish and other creatures it eats by spearing them with the point? That hardly seems likely, because female narwhals eat the same things males do, and they don't have spears to catch them with. Narwhals have been seen using their tusks to stir up sand on the sea bottom and drive shrimp out of hiding, but most scientists doubt this is the real pur-

BELUGA OR
WHITE WHALE

NARWHALS

Beluga or White Whale

pose of the tusk. For, once again, if the tusk is a tool for getting food more easily, females would have it, too, just as male and female elephants both have trunks. Some scientists think male narwhals may use their tusks in fights over females, but most scientists believe the tusks are probably just ornaments that only the males have, just as only a male chicken (a rooster) has a comb and only a male lion has a mane.

At one time, narwhal tusks were as precious and costly as rare jewels! In Europe, during the Middle Ages, people believed that a narwhal tusk was actually the horn of a fabulous, magical beast known as a unicorn—a small, white horse that had a horn growing out of its forehead. A unicorn's horn was supposed to be able to make any poison harmless, so European kings (who were always fearful of being poisoned by someone who wanted to be king instead of them) were anxious to have a drinking cup made of "unicorn horn." Viking traders from the North brought Narwhal tusks into Europe and got rich from them. One German ruler once paid 100,000 silver coins for a single narwhal tusk!

The other member of the tiny family the narwhal belongs to is the white whale—which is also called the beluga, from a Russian word that means "white." Adult white whales are a yellowish-white color, but oddly enough, the whales are born brown, then become dark gray, then yellowish, and only turn white when they are four or five years old.

White whales are about the same size as narwhals. They have stout bodies, blunt noses, and each has a noticeable bump, or melon, on top of the head. They also each have a row of little bumps on their back, rather than a fin. Unlike narwhals, white whales are rather slow swimmers—one of them in a real hurry can't move much faster than about 11 miles an hour, which is only about half as fast as you can probably run. And, most of the time, a white whale moves at only half that speed.

White whales live where narwhals do, in the Arctic Ocean. However, they seem to prefer the water close to the coasts of land —North America, northern Europe, and Russia. They live in little groups of five or ten, although at times as many as a hundred may be seen moving through the water together. But even though there are many of them, they are hard to see because they blend right in among the many chunks of yellowish-white ice that dot the Arctic Ocean in the summertime.

Unlike narwhals, white whales have a number of teeth in both their upper and lower jaws—from 32 to 40 teeth altogether. However, they, too, swallow all their food whole. They do their hunting in shallow water near coasts and eat the same things narwhals do: fish, squid, crabs, and shrimp.

Long ago, British sailors aboard whaling ships gave white whales the nickname of Sea Canaries, because these whales make whistling sounds that are much like the song of a bird. Even though the whistles are made underwater, they are loud enough to be heard above the surface. The whales whistle through their blowholes, as dolphins do.

Even though the Arctic is their home, its fierce cold can cause problems for narwhals and white whales. In winter, much of the sea freezes over and is covered by ice that is as much as 10 feet thick in places. The whales must move away from these regions of thick ice, for they would not be able to put their heads through it in order to breathe and would drown. They generally move south in winter, and narwhals and white whales have been seen as far south as the waters off Alaska, Scotland, and Norway. And once, in 1966, a white whale came all the way down to northern Germany and lived for a month in the Rhine River.

However, groups of whales at times get trapped in bays with the water slowly freezing around them. Then they are in serious trouble. As the water freezes, they must keep breaking through the ice by ramming it with their heads, but if the ice freezes too thickly, they are doomed.

There is another danger, too. Often, when large numbers of narwhals have become trapped in places where there was only a small area of unfrozen water, they have been slaughtered by Eskimos as they crowded up into the water for breaths of air. For thousands of years, narwhal's skin, which is rich in vitamins, has been a precious food for these people of the North. As many as 200 narwhals at a time, trapped in a small pool of water in the ice, have been killed by Eskimos seeking this food that was necessary for *their* survival.

The Beaked Whales—A BIG-NOSED FAMILY

THE BEAKED WHALES are a big family of toothed whales whose jaws form beaks. There are about 18 different kinds, many of which are rather rare. They all eat mostly squid and some fish.

Most male beaked whales have only two teeth, both in the lower jaw. Some kinds of these whales have their teeth at the front of the jaw, and some have them back toward the middle so they look like fangs sticking up on each side of the whale's mouth! Female beaked whales usually have no teeth showing; their teeth are down inside the gum. Inasmuch as the females have no visible teeth, and the bodies of some kinds of older males are often covered with long scars, it seems as if the males sometimes use their teeth to do battle with each other.

Most kinds of beaked whales have teeth that are rather small and pointed. But one kind, called the strap-toothed whale, has large, flat teeth that curl up over its upper jaw in such a way that it can hardly open its mouth! The strap-toothed whale is about 16 feet long and lives in the South seas.

The best known of the beaked whales is the stout bottle-nosed whale, which is about 32 feet long. Bottle-nosed whales live in the North Atlantic in herds of from ten to a hundred. A large herd is usually led by an old male, which can be told from younger whales by its color. As a bottle-nosed whale ages, its color changes from chocolate-brown to whitish yellow.

Bottle-nosed whales are apparently the champion "breath holders" of all the whales. Fishermen have seen some of these creatures dive down and stay underwater for as long as two hours. They can go down more than a quarter of a mile.

The beaked whales called goose-beak whales, which may be as much as 23 feet long, stay in warm water near the tropics. A group of these whales can be a puzzling sight because all the whales in the group may be differently colored! Goose-beak whales can be dark gray, rusty brown, or two-toned tan, and may be polka-dotted with whitish splotches. Old males have white heads, like gray-haired old men.

The largest of the beaked whales is Baird's beaked whale, also called the giant bottle-nosed whale. It may reach a length of 42 feet, which makes it the second largest of all toothed whales, after the big sperm whale. Unlike most other beaked whales, Baird's beaked whale has four teeth at the front of its lower jaw.

Bottle-nosed Whale

52

Baird's Beaked Whale

BAIRD'S BEAKED WHALES

Whale Babies

A FEMALE bottle-nosed dolphin, swimming with a group of other dolphins in water near the Florida coast, began to lag behind the others. She moved slowly, for her body was suddenly behaving strangely to her. She felt quick little jabs of pain and a sensation of movement within her stomach. Her time had come to give birth to a baby.

Several of the other female dolphins of the group also slowed down. They dropped back close to the slow-moving female and swam beside her protectively. They seemed to sense what was about to happen.

After a time, the female's stomach muscles gave a sudden squeeze. The baby was beginning to come out. The tip of its little tail had pushed through the opening in the mother's body.

Slowly, more and more of the little dolphin appeared—its tail, then its middle part, then its small flippers. All the while, the other female dolphins continued to swim around the mother, like guards.

Finally, after about three-quarters of an hour, the baby's head emerged. It was born! The long tube that had connected it to its mother, through which it had received air and nourishment while it was in her body, broke off. At that moment, the baby

dolphin became ready for life on its own!

At once, it swam up to the surface of the water. Its blowhole popped into the air and the baby took its very first breath. Its mother was right beneath it, so close that they were nearly touching. Had the baby faltered on its way to the surface, the mother would have pushed it above the water with her body.

After a time, the mother picked up speed, swimming faster, to catch up with the rest of the group. But the baby wasn't left behind, nor did it have to work hard to keep up with her. Still so close to her that their bodies were practically touching, it was actually carried right along with her by the force of the water streaming past them.

In about an hour, the mother slowed down and rose to the surface. She seemed to feel the time had come to give her little one its first meal. She rolled over and began swimming slowly on her side. By doing this she made sure that when the baby pushed its nose against her belly to feed, its blowhole would be above water, so it could breathe.

The baby wasn't quite sure what to do, but instinct made it shove its mouth toward

the right spot. After a moment, it found a nipple. Rich, warm milk squirted into its mouth. Contentedly, the baby fed.

Then, as most young mammals do after eating, the little dolphin took a nap. Eyes closed, it was carried along again with its mother as she sped through the water. But, the nap was a short one, only a few minutes long. The baby dolphin was much too alert and interested in life to sleep much!

Whales, like humans, usually have only one baby. However, a whale mother will sometimes have twins, just as a human mother does. And, once in a very great while, whale triplets are born. But there doesn't seem ever to have been any quadruplets or quintuplets, such as a few human mothers have had.

The birth of a baby dolphin is very different from the birth of most kinds of land mammals. Baby humans, dogs, cats, horses, and most other land mammals come out of their mother's body headfirst when they are born. But baby dolphins, and apparently all other kinds of toothed whales, come out tailfirst. There seems to be a good reason for such a difference. Whales are air breathers, and they breathe through nostrils in their head. It sometimes takes a baby dolphin as long as an hour to get born, so if it came out headfirst it could *drown* while the rest of it was coming out. However, scientists have found that baleen whales are apparently born headfirst, just as most land mammals are—which is rather puzzling. There are some mysteries about the way whales are born that remain to be solved.

What must life seem like to a newborn baby whale? Far, far different than for a baby land mammal. A newborn human, dog, or cat is weak and helpless. It can barely move, except to feebly jerk its limbs. Many newborn mammals cannot even see at first. But a baby whale born on the broad, deep ocean is strong, able to see, and able to swim the instant it leaves its mother's body—and it has to be! A baby whale has no warm, safe den to stay in while it grows bigger and stronger, as a baby bear has. A little whale can't snuggle in the safety of a pouch on its mother's body, as a baby kangaroo can. A newborn whale is on its own in an immense expanse of water where there is no place to hide and where there are enemies such as killer whales and sharks—to which a baby whale is an easy, tasty meal!

However, baby whales are apparently well protected. As old-time whalers knew, mother whales will fight furiously to protect their young ones. Among many kinds of whales, a mother and another female will often team up to protect a youngster. If the baby seems to be in danger, its mother or the other female puts herself between the young one and whatever is threatening it. The female, which scientists refer to as an "aunt," will often act as a rear guard, blocking the path to the mother and baby so they can swim away to safety.

Even young whales that have grown up enough to be on their own can count on help if they get into trouble. A young sperm whale once got curious about a small boat and, while nosing too near the boat's whirl-

ing propeller, got itself badly slashed. The youngster began to make noises that were apparently cries for help, for within half an hour 27 female sperm whales had come, from all directions, to help the little whale. Some of them must have come from many miles away.

Whale mothers are just like human mothers in many ways. For example, when a human baby starts to eat solid foods its mother is usually very careful to make sure the food is in small bits and free of any bones or seeds that might make the child choke. A mother dolphin does much the same sort of thing. When a baby dolphin first begins to eat fish, its mother will break off the heads for it, so it won't be in danger of choking on the hard skull bones.

Whale mothers often play with their babies, just as human mothers do. Of course, it's a watery sort of play. A gray whale mother will dive just beneath her baby and squirt a blast of air up out of her blowhole, causing the obviously delighted little whale to spin around and around in a whirl of bubbles! Then, the mother will rise up and, with gentle bumps of her nose, make the young one bounce up and down.

And, whale mothers often have to discipline their babies, just as human mothers must sometimes do. For, whale babies seem to enjoy bumping mama with their noses, crawling up onto her when she's trying to doze at the surface of the water, and other mischievous and annoying things! A mother right whale will grab her youngster with her flippers and hold it until it calms down. A mother dolphin will give her young one a nip or will hold it underwater for a few moments. The baby will wiggle and make noises of protest, but after that it will behave itself—for a little while, at least.

Most kinds of whale mothers nurse their babies (that is, they continue to feed them milk) for ten months to a year. Blue whales and other rorquals nurse their babies for about five to seven months. People who have tried whale milk say it tastes like a mixture of oil, liver, fish, and milk of magnesia. But it is a rich, nourishing food, and baby whales thrive on it. A baby blue whale puts on about $8\frac{1}{2}$ pounds an *hour* during the six or seven months while it is drinking its mother's milk. It drinks about 130 gallons a day! Of course, a blue whale baby is absolutely the *biggest* baby in the world —as much as 25 feet long when it is born— so it needs a lot of food.

A young whale stays very close to its mother during the months she is nursing it. Then, when it begins to eat krill or fish or other grown-up food, it spends less and less time near her. Now it knows how to get its own food and how to take care of itself. In time, it goes off on its own. By the time it is from four to six years old, it has become an adult, ready to have babies of its own.

Present-day Factory Ship

The Future of Whales

MANY KINDS OF whales were ruthlessly hunted for hundreds of years, and their numbers grew fewer and fewer. Whereas there had once been at least 100,000 of the playful, singing humpback whales, there were no more than 7,000 left by the end of the 1970s. There were only about 13,000 of the giant blue whales left out of some 200,000. At this moment, there are probably no more than about 4,000 black right whales and probably only about 2,500 Greenland right whales.

By the end of the 1970s, the United States, Great Britain, and many other nations that once did a lot of whaling, no longer allowed any of their people to hunt whales. Only two nations, Japan and Russia, still had fleets of whaling ships. There were also some "pirate" whaling ships that didn't officially belong to any country, but that hunted whales on their own and sold the meat and oil wherever they could.

Many nations, even those that still allowed whale hunting, belonged to an organization called the International Whaling Commission. This organization decided which kinds of whales were still plentiful enough to be hunted and set the amount of them that could be killed each year. It also decided which kinds of whales should be left alone. Blue whales, right whales, gray whales, and humpbacks were officially protected by the International Whaling Commission, which urged all whalers not to harm any of them. However, the commission had no police force to make the whalers obey its rules, and many of these protected whales were still killed each year by the pirates, who paid no attention to the commission's rules.

Many people throughout the world felt that all whale hunting should be stopped. They knew that the laws protecting whales were useless if the pirates and the nations that still allowed whaling chose to ignore them.

But, many other people felt that whale hunting had to go on. For millions of people in Japan and a few other countries, whale meat was an important food. So, while some members of the International Whaling Commission, such as the United States and Great Britain, insisted that whaling should stop, others, such as Japan and Russia, insisted it had to keep on going.

However, in 1979, at a meeting of all the members of the International Whaling Commission, some rules were passed that

Modern Whaler

many members hoped might finally bring an end to whale hunting. For one thing, Japan promised that it would no longer buy whale meat from pirate ships—which meant the pirates could no longer make any money by killing whales and might as well stop. For another thing, all the members agreed that no more whale hunting would be done in the Indian Ocean—instead, it would become a place where whales could live in safety, a whale refuge. And a third, very important thing was that deep-sea factory ships, which could search out whales anywhere in the world, killing them and turning them into canned meat and oil in a few hours, could no longer be used by the countries that still allowed whaling. This meant that whales could only be hunted with ordinary ships, which is a much more expensive way to hunt them.

So, although Japan and Russia still had whaling fleets in 1979, most people felt it had become so expensive and difficult to hunt whales that these two nations would soon have to stop. Perhaps they will have stopped by the time you read this book.

But, even if that happens, whales are still not completely safe. Some kinds of them are in serious danger. There simply may not be enough Greenland right whales, black right whales, and humpback whales to keep their races going. These kinds of whales may grow fewer and fewer—until none are left.

And even the whales that are still plentiful now may someday be in trouble. The number of people in the world is growing tremendously. Perhaps soon most of the fish, and even the *krill* that whales eat, may have to be taken from the sea to help feed people. If humans take most of their food, whales could soon die out.

There is also the problem of pollution. The sea is polluted with oil, spilled from ships and undersea wells, and with chemicals that have run into the water from factories and farms on land. This has already caused trouble for some sea creatures—it may cause trouble in the future for whales.

So, the future of whales is still rather uncertain. Everyone who wants to see these big creatures saved—to live as they did before humans began to hunt them—can only hope that we will be able to solve problems such as pollution and overpopulation without causing any more problems for our big cousins in the sea.

Index

PRINTED IN U.S.A.

Pronunciation Guide

albino	ahl-BY-noh
Amazon (dolphin)	AM-uh-zahn
ambergris	AM-buhr-gris
Baird's (beaked whale)	BAHRDZ
baleen	buh-LEEN
barnacle	BAHR-ni-kuhl
beluga	buh-LOO-guh
bowhead	BOH-hed
Bryde's (whale)	BROO-dahz
diatom	DY-uh-tahm
dolphin	DAWL-fuhn
fluke	FLOOK
Ganges (River dolphin)	GAN-jeez
Indus (River dolphin)	IN-duhs
kraken	KRAHK-uhn
lamprey	LAM-pray
mammae	MAM-ee
Minke (whale)	MING-kee
narwhal	NAHR-wahl
plankton	PLANG-tuhn
porpoise	PAWR-puhs
pygmy (whale)	PIG-mee
rorqual (whale)	RAWR-kwuhl
sei (whale)	SAY
sonar	SO-nahr
spermaceti	spuhr-muh-SET-ee